Copyright © by Harcourt, Inc.

All rights reserved. No part of this publication may be reproduced or transmitted in any form or by any means, electronic or mechanical, including photocopy, recording, or any information storage and retrieval system, without permission in writing from the publisher.

Requests for permission to make copies of any part of the work should be addressed to School Permissions and Copyrights, Harcourt, Inc., 6277 Sea Harbor Drive, Orlando, Florida 32887-6777. Fax: 407-345-2418.

HARCOURT and the Harcourt Logo are trademarks of Harcourt, Inc., registered in the United States of America and/or other jurisdictions.

Printed in Mexico

ISBN-13: 978-0-15-362258-8
ISBN-10: 0-15-362258-X

2 3 4 5 6 7 8 9 10   050   16 15 14 13 12 11 10 09 08

Visit The Learning Site!
www.harcourtschool.com

What Are Forces?

Why are some things harder to push or pull than others? Every day, you use pushes and pulls to get things done. Sometimes you use a big push, such as when you push open a heavy door. Other times you use a small pull, such as when you pick up a pillow. The heavier an object is, the harder you must push or pull to get it to move. All pushes and pulls are forces. A **force** is a push or pull that can cause a change in motion. A force may start things moving, speed them up, slow them down, or make things change direction.

Suppose you kick a soccer ball. What happens? You've applied a force to the ball—it starts to move! Eventually, the ball slows down and stops. Why? Another force is acting on the ball. This force is called friction. **Friction** makes things either slow down or stop. It can even keep things from moving at all.

Fast Fact

Friction makes heat when surfaces rub together. You can test this quite easily. Try rubbing your hands together quickly. What happens? They get warmer!

Nothing moves unless a force acts on it. A push from a foot set this ball in motion.

What causes friction? Friction occurs when one surface rubs against another. Different materials produce different amounts of friction. Rough materials rub more and create more friction. Most smooth materials do not rub as much as rough materials. Smooth materials produce less friction. Some materials, such as rubber, are smooth but still produce a lot of friction.

 CAUSE AND EFFECT Picture a ball lying on the sidewalk. What must you do to make the ball move?

When you ride a bike, friction is present between the bike's tires and the road.

Familiar Forces

Even though all forces push or pull, there are different kinds of forces. You have already learned about friction, the force present whenever surfaces touch. But some forces push and pull objects without touching them at all. One of those forces is acting on you right now. The force of gravity is pulling on you and everything else on Earth all the time.

Gravity is the force of attraction between objects. Every object in the universe pulls on every other object with **gravitational force**. This force can be either weak or strong. The force of gravity between two objects depends on their mass and how far apart they are. The closer and larger the objects are, the stronger the gravitational force between them. Most objects do not have enough mass for you to notice their pull on each other. Earth has a lot of mass, so the pull between it and other objects is strong.

The gravitational pull of Earth on the fish is shown on the spring scale.

A magnet has two ends, called *poles*. The force a magnet exerts is strongest at its poles.

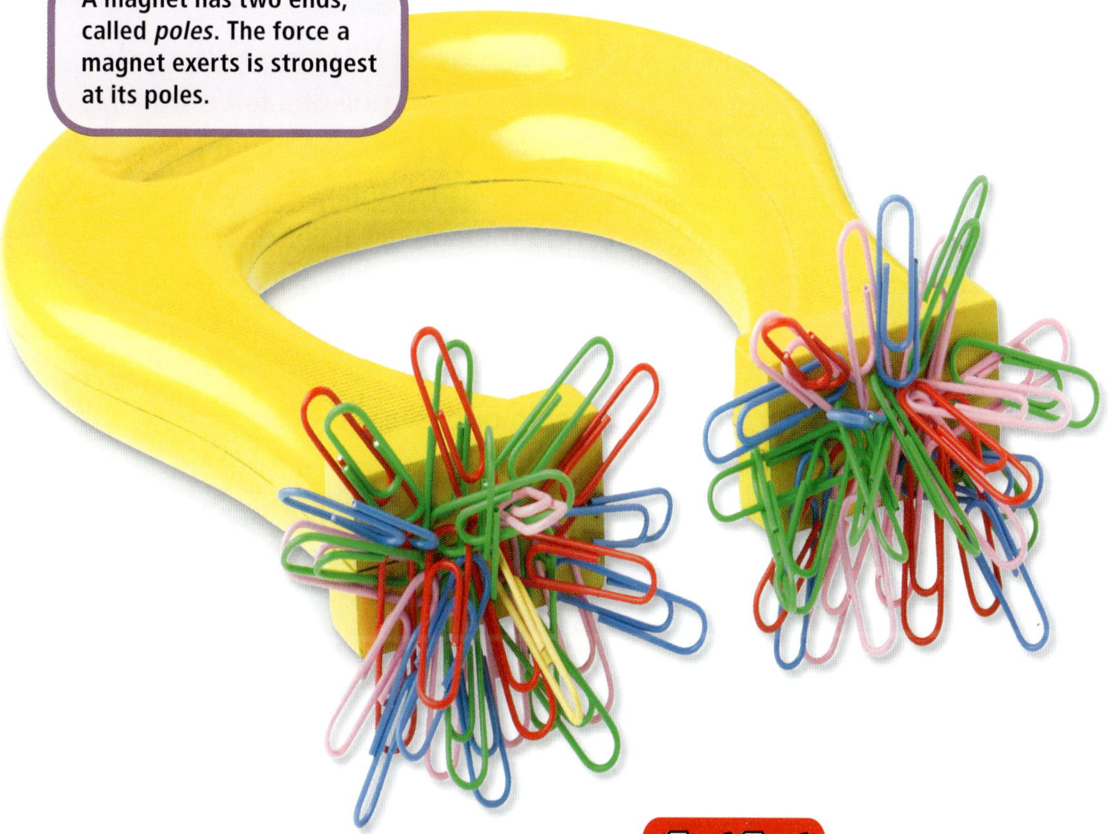

Another familiar force is magnetism. **Magnetism** is the force produced by a magnet. What happens when you hold a magnet near an object made of iron or steel? The magnet pulls on the objects, and they stick to the magnet! Many devices, such as computer disks, electric motors, and cassette tapes, work because of magnetism.

Fast Fact

The needle of a compass is a magnet. The compass needle points along an imaginary line connecting the North and South Poles. Why? Because Earth itself is like a giant magnet!

CAUSE AND EFFECT How does the force of gravity make some of the things you do each day harder? How does it make some things easier?

Staying Balanced

Did you know that forces work in pairs? Whenever you push or pull on something, you feel the push or pull of a force working in the opposite direction. Picture yourself walking. Your feet push against the floor, but the floor also pushes against your feet!

Sometimes the forces acting on an object balance each other. For example, suppose two people are sitting on a seesaw. If the weights of the people on each end of the seesaw are the same, what happens? Nothing! The forces cancel each other out. They are equally strong, but act in opposite directions. The seesaw doesn't move.

When forces acting on an object cancel each other out, they are called **balanced forces**. When balanced forces act on an object, they do not change the object's motion or direction. This means if the object is stopped, it will stay stopped. If moving, it will keep moving at the same speed and direction. Because the forces are balanced, it seems as if no force is acting on the object at all.

The forces acting on each side of the scale are balanced.

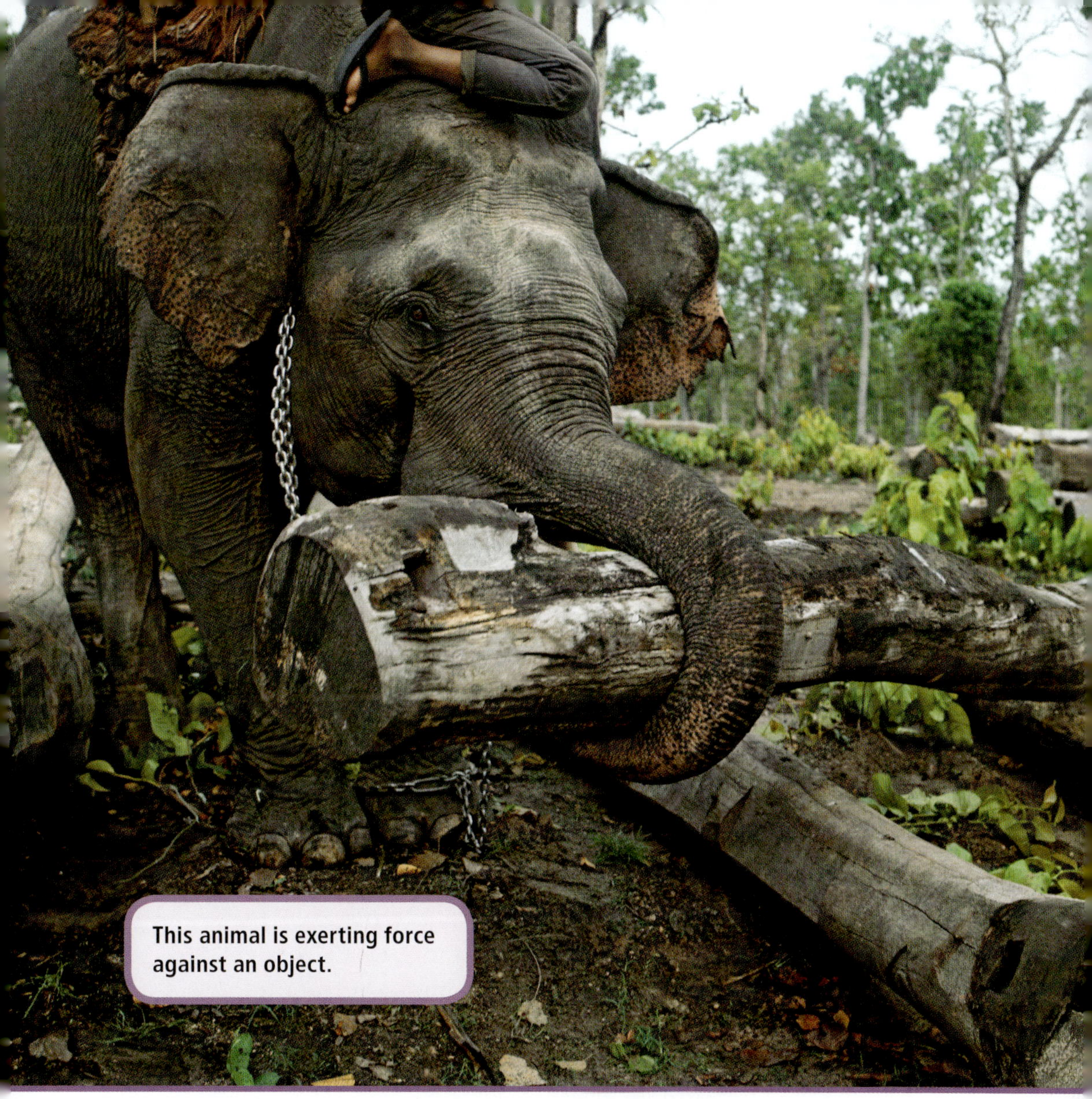

This animal is exerting force against an object.

Suppose you exert a force by pushing on a very heavy object, such as a piece of furniture. The object doesn't move because it is exerting an opposite force that balances the force you exert. Even though the object doesn't move, you are still exerting a force.

Focus Skill **COMPARE AND CONTRAST** **Two teams are playing tug-of-war, but neither team is winning. How do the forces acting on the teams compare?**

When Forces Are Unbalanced

When forces acting on an object cause a change in motion, they are called **unbalanced forces**. The changes may cause the object to start to move, change direction, speed up, slow down, or stop. Suppose you push against a sturdy object, such as a wall. Although you exert a force, no movement occurs because the wall is pushing back with an opposite force that exactly balances the force you exert. The forces are balanced, so they cancel each other out. Suppose a bulldozer pushes against the same wall. It may exert a force large enough to overcome the force exerted by the wall. What happens? The wall falls down!

The force exerted by the bulldozer is greater than the force exerted by the dirt. Unbalanced forces make the soil move.

Unbalanced forces can cause an object not only to move but also to stop. For example, when you catch a ball, you stop it by exerting a force greater than the force exerted by the moving ball. Unbalanced forces can also speed up or slow down a moving object. A person paddling a boat upstream may be slowed by the force of the current, even though he or she exerts a strong force on the water. If the current is strong enough, its force becomes the greater force, and the boat goes backwards.

 COMPARE AND CONTRAST The forces on a child sitting at the top of a slide are balanced. Contrast the forces in this situation to the forces when the child starts to move.

> A person in a boat may have a hard time paddling against the force of this fast-moving water.

What Is Work?

Are you working right now? All day long you do different kinds of work, such as schoolwork and homework. You may have chores that you do, too. What do you think work is?

When people talk about work, they usually mean a job that adults do. Whether or not you have a job, you work every day! You work when you pedal a bike. You work when you play ball or jump rope. You might think you work when you read a book. In science, the only way to do work is to make something move. Scientists have a precise definition of work. They say that **work** is done when a force moves an object through a distance.

Animals work, too! This ant is exerting a force to move the leaf.

In building this nest, this bird is doing a lot of work. Think of all the objects moved through a distance!

Suppose a gardener is trying to pry a large rock out of the ground. The gardener pushes and pulls, but the rock doesn't budge. Now suppose that in a tree nearby, a squirrel picks up an acorn. Who has done more work, the gardener or the squirrel? According to the way scientists define work, the gardener has not done any work at all. Although the gardener has exerted a large force, the gardener did not move an object through a distance. The squirrel exerted a much smaller force. But the squirrel has done work, because an object—the acorn—was moved.

Fast Fact

Work is measured in units called *joules*. One joule (J) is equal to a force of 1 newton (N) exerted over a distance of 1 meter (m). For example, a medium-sized apple weighs about 1 N. If you lift that apple 1 m, you do 1 J of work.

 MAIN IDEA AND DETAILS You push against the wall of a building as hard as you can. A friend picks up a paper clip. Who did work? Explain.

Work and Machines

You have learned that work is the product of the force applied to an object and the distance it is moved. Here is an equation that shows the definition of work: Work = Force x Distance.

Trying to lift heavy objects over a long distance can be a lot of work. A machine can make the work easier. A **simple machine** may change the size or direction of a force, or the distance over which the force acts. Simple machines don't actually reduce the amount of work, but they make it easier.

One example of a simple machine is a lever. A **lever** is a bar used to make it easier to move things. The two parts of a lever are the lever arm, which moves, and the **fulcrum**, or balance point, which does not move.

This crowbar works as a lever to remove nails or pry things apart.

A **wheel-and-axle**, another simple machine, is made up of a wheel with a rod, the axle, running through the center. Like a lever, a wheel-and-axle makes work easier by using less force over a greater distance. Doorknobs and steering wheels are examples of wheel-and-axles.

 MAIN IDEA AND DETAILS A certain simple machine reduces the amount of force needed to move an object. What must it do to the distance through which the force is exerted?

A wheel-and-axle type of simple machine is part of this fishing rod.

13

More About Machines

Another simple machine is the pulley. A **pulley** is a specialized wheel with a groove to hold a rope. Pulleys make work easier by changing the direction of the force needed to move an object. You can pull down to lift something up!

An inclined plane is also a simple machine. An **inclined plane** is a sloping surface, like a ramp. Using a ramp means going a longer distance, but doing it with less effort. A wedge is a movable inclined plane. Axes and plows are examples of wedges.

Most of the machines you use every day are compound machines, machines made up of two or more simple machines. For example, a shovel is a wedge attached to the end of a lever. A hand-cranked pencil sharpener uses a wheel-and-axle to turn a set of wedges that cut. What other compound machines have you used?

 MAIN IDEA AND DETAILS How does an inclined plane make work easier?

A can opener is a compound machine. It is made up of a wedge, two levers, and a wheel-and-axle.

This ramp is an example of an inclined plane.

Summary

A force is a push or a pull that can change the motion of an object. You see the effects of friction, gravity, and magnetism each day. When the forces acting on an object are balanced, no change in motion occurs. When the forces acting on an object are unbalanced, there is a change in motion. Work occurs when an object is moved. Simple machines make work easier.

Glossary

balanced forces (BAL•uhnst FAWRS•iz) Forces that act on an object but cancel each other out (6)

force (FAWRS) A push or pull that may cause an object to move, stop, or change direction (2, 4, 5, 6, 7, 8, 9, 10, 11, 12, 14, 15)

friction (FRIK•shuhn) A force that opposes motion (2, 3, 15)

fulcrum (FUHL•kruhm) The balance point on a lever that supports the arm but does not move (12)

gravitational force (grav•ih•TAY•shuhn•uhl FAWRS) The pull of all objects in the universe on one another (4)

gravity (GRAV•ih•tee) The force of attraction between objects (4, 5, 15)

inclined plane (in•KLYND playn) A ramp or another sloping surface (14, 15)

lever (LEV•er) A bar that makes it easier to move things (12, 13)

magnetism (MAG•nuh•tiz•uhm) The force produced by a magnet (5, 15)

pulley (PUHL•ee) A wheel with a rope that lets you change the direction in which you move an object (14)

simple machine (SIM•puhl muh•SHEEN) A device that makes a task easier by changing the size or direction of a force or the distance over which the force acts (12, 14, 15)

unbalanced forces (uhn•BAL•uhnst FAWRS•iz) Forces that act on an object and don't cancel each other out; unbalanced forces cause a change in motion (8, 9)

wheel-and-axle (weel•and•AK•suhl) A wheel with a rod, or axle, in the center (13, 14)

work (WERK) The use of a force to move an object through a distance (10, 11, 12, 13, 14, 15)